YOUR KNOWLEDGE HAS VALUE

Prasad Mp

Micobial degradation of triphenyl methane dyes

GRIN Publishing

Bibliographic information published by the German National Library:

The German National Library lists this publication in the National Bibliography; detailed bibliographic data are available on the Internet at http://dnb.dnb.de .

Imprint:

Copyright © 2012 GRIN Verlag GmbH
Print and binding: Books on Demand GmbH, Norderstedt Germany
ISBN: 978-3-656-86711-1

This book at GRIN:

http://www.grin.com/en/e-book/286443/micobial-degradation-of-triphenyl-methane-dyes

GRIN - Your knowledge has value

Since its foundation in 1998, GRIN has specialized in publishing academic texts by students, college teachers and other academics as e-book and printed book. The website www.grin.com is an ideal platform for presenting term papers, final papers, scientific essays, dissertations and specialist books.

Visit us on the internet:

http://www.grin.com/

http://www.facebook.com/grincom

http://www.twitter.com/grin_com

MICROBIAL DEGRADATION OF TRIPHENYL METHANE DYES

Abstract

Try phenyl methane dyes have been found in soil and river sediments as consequences of improper chemical waste disposal. 10000 dyes and pigments are produced annually world wide amounting to 7*107tones which are hazardous and pose serious environmental problems. It is estimated that 10-15% of the dye is lost in the effluent during the dying process. Try phenyl methane dye decolorizing bacteria have been isolated; there are few reports of specific enzymes that decolorize these dyes. Isolate bacterial strains which had the capability to decolorize textile dye like bromophenol blue, crystal violet and malacate green. We estimate the decolorization percentage for all the three tri phenyl methane dyes and quantify the activity of the TMR enzyme that degrades try phenyl methane dye and characterize the dye degrading organism as *pseudomonas* species.

Keywords: Try phenyl methane, TMR enzyme, Dyes and *pseudomonas* species.

Introduction:

Dyes are coloring pigments that impart color to the substrate when they are in solution form. Technically dyes are distinguished from the intermediates based on the presence of auxochrome, the group that allows the basic unit to attach and impart color to the substrate. Dyes are derived synthetically from raw materials like hydrocarbons, benzene, toluene, naphthalene and anthracene using coal tar obtained from distillation of coal (Alinsafi A, et al., 2006). Both organic and inorganic materials are needed to make dyes and intermediates. The raw material sequence for making dyes is petroleum hydrocarbons intermediates dyes (Bell J, et al., 2000). N

1

Dyes are retained in substrates by physical absorption, metal complex formation or by the formulation of covalent chemical bonds and they obtain their color due to electronic transitions between various molecular orbital where intensity of the color is determined by the probability of transitions (Cariell, C.M, et al., 1996). Dye is a substance (generally an organic compound), which is used for imparting permanent color to textiles - silk, wool and other substances. There are two types of dyes. Natural dyes occur in nature e.g. Indigo (a blue dye), alizarin (a red dye).Synthetic dyes are man-made dyes e.g. Crystal violet (a bluish green dye), azo dye, aniline yellow, orange etc. A colored substance can act as a dye only if it can be fixed to the material being dyed (Daneshvar N, et al., 2007). At the same time it should be resistant to the action of light, water and soap. The important condition for a colored compound to act as a dye is the Presence of chromophore. These are the groups, which are responsible for producing a color to a dye because they are capable of absorbing light in the ultra violet region. Some important chromophores are: -N=O, -N=N-, -C=N, (CH=CH). A chromogen without auxochrome can never act as a dye. 10,000 dyes and pigments are produced annually worldwide amounting to 7×10^5 tones which are hazardous and pose serious environmental problems. It is estimated that 10-15% of the dye is lost in the effluent during the dying process (Lacalle M, et al., 2007).

The recent high profile of color pollution is mainly the result of increasing public awareness and expectations of the environment; coinciding with rising levels of color discharges (Oxspring DA, et al., 1996). One of the more pressing environmental problems that have been facing the textile industry is the removal of the color from dye bath effluent prior to discharge to local sewerage treatment facilities or adjoining watercourses. Considerable efforts have been made on developing suitable treatment systems for these effluents. Wastewaters originating from reactive dye processes have created a particular problem because the dyes can exhibit low levels

2

of fixation with the fiber. The brightly colored unfixed dyes are highly water-soluble and are not removed by conventional treatment systems (Robinson T, et al., 2001). This is particularly noticeable as the human eye can detect reactive dyes at a concentration as low as 0.005 mg/l in clear waters. Discoloration of textile dye effluent does not occur when treated aerobically by municipal sewerage systems. Though the formation in 1974 of the Ecological and Toxicological Association of the Dye stuff Manufacturing Industry (ETAD), aims were established to minimize environmental damage, protect users and consumers and to cooperate fully with Govt. and public concerns over the toxicological impact of their products. Over 90% of some 4000 dyes tested in an ETAD survey had LD50 values greater than 2×10^3 mg /kg. The highest rates of toxicity were found amongst basic and diazo direct dyes .

Materials and Methods

Isolation of the dye degrading organism

Soil is collected from the textile industries in and around Hyderabad. This soil is then serially diluted for isolation of bacterial colonies. The 10^{-5} diluted soil sample is and spread over nutrient agar media along with a dye (bromophenol blue, or malachite green)with concentrations of 20 μM, 40 μM, 60 μM,80 μM. Whereas for crystal violet the concentrations ranged from 20 μM,30 μM ,40 μM ,50 μM The petri plates are incubated for 48 hrs at 37^0C. The colony which has shown maximum clearing zone (≈10mm) was selected and maintained as pure culture in nutrient broth.

Decolorization assay

For degradation experiments, the pure cultures were inoculated into flasks at 4% (v/v) level. All experiments were conducted in the same conditions consisting of a 500-mL flask containing 250 mL nutrient broth (pH 7.5). Culture medium was supplemented with 50 mg/L of the dye to be tested in the presence of yeast extract (0.1%) and glucose (7 mM). Flasks were incubated at 37°C, under shaking (120 rpm) in a rotary shaker. Samples were collected at different time points (2, 4, 6, 8, and 24 h) to determine the dyes decolorization.

Decolorization Assay

For determination of CV, BB and MG color removal, 5 ml aliquots of the culture were sampled at different culture periods (2, 4, 6, 8 and 24 h), centrifuged at 6,000 rpm for 10 min to eliminate the bacterial cells, and the supernatant was examined by spectrophotometer at the (λmax) of 590 nm, 600 nm and 618 nm, respectively, for CV, BB and MG. The percentage of decolonization was calculated as following:

Decolonization (%) = [(Absorbance at t_0) − (Absorbance at t_1)]/ (Absorbance at t_0) × 100]. Dye elimination was investigated with each selected bacterial isolate of that particular dye.

100 ml nutrient broth was prepared and poured into four conical flasks (25 ml each). Different concentrations of bromophenol blue (20μM, 40 μM, 60 μM, 80 μM) were added to each flask. Inoculate a 1 ml of culture from the nutrient broth and incubate the flasks at 37°c. Observe the OD values after 24 hrs of incubation at 600 nm. Similarly for each concentration of the dye, control is set up without the culture.

Depolarization % =O.D (control − test)/ O.D of control*100

4

TMR Enzyme assay

Since the dyes are triphenyl methane dyes and the organism is able to reduce the dye, therefore triphenyl methane reductase enzyme assay was done. The standard assay system for TMR comprises 20mM sodium phosphate buffer (pH 7), 20µM dye (bromophenol blue), 0.1mM NADH and a suitable amount of culture broth in a total volume of about 1ml. This is incubated at room temperature for about 2 mins and then OD is observed at 600nm.

The enzyme activity is calculated as follows

In general terms the enzyme activity (µmol /min/ml) = ΔAx 1000/ extinction coefficient of bromophenol blue x vol in cuvette x 1.0/vol used for assay.

Extinction coefficient of bromophenol blue = 60,500

Extinction coefficient of crystal violet =111000

Extinction coefficient of malachite green=137000

The final activity is obtained by multiplying by any dilution factor for micromoles per min per ml and dividing by the protein concentration in mg per ml for micromoles per min per mg.

Characterization of the bacterial isolates

Take a clean glass slide and place a drop of water over it. Now, sterilize the inoculating loop and pick a part of colony and spread over the water drop, make thin smears of it. Let the smears air dry and then heat fix. Stain the smear with crystal violet for 1-2 minutes, then wash it with distilled water, now cover the smear with grams iodine for about 1 min and wash the slide

5

with distilled water, then add acetone (alcohol) drop by drop for about 30secs and immediately wash with distilled water. Lastly add saffranin to the smear for 2 minutes and wash with distilled water. Blots dry the slide using a blotting paper. Let the slide dry completely and then observe the fluid under 10x, then observe morphology under 45x and finally report the gram morphology and arrangement under oil immersion lens (100x).

IMViC tests

To identify the three isolated organisms IMViC (Indole production, methyl red reduction, Voges Proskauer and citrate utilization) tests are done for each of the culture.

Citrate

Prepare Simmons citrate agar media 100ml. Citrate media contains a carbon source of sodium citrate, a nitrogen source of ammonia and a pH indicator of bromothymol blue. Slants of citrate agar were made and the organism is inoculated and incubated at 37° for 24hrs. If the agar turns blue, then it is positive for citrate test.

Oxidase test

Oxidase reagent: N,N tetra methyl para phenyl diaminedihydro chloride

Whatman filter paper

1%oxidase reagent with sodium thoinate

Pick up a 24hr old culture and smear the culture on Wattmann filter paper and then in pregnant with 2-3 drops of 1%oxidase reagent reduced with sodium thoinate. Wait for 10 secs. If purple colour appears within 10 secs, it is a positive test for oxidase. But if the colour produses between 10-60 secs, it is delayed positive reaction while absence of colour is a negative reaction.

Estimation of DNA by DPA (Diphenyl amine) method

A liquid sample of standard DNA whose concentration is to be estimated is taken. The volume is made to 1ml by adding distilled water. 4ml of DPA is added to all the tubes. The contents of the tube are mixed thoroughly and heated for 10 mins in a water bath. The solution in the tubes turns blue colour. Now the tubes are cooled and the OD is read at 600nm.

Amplification of TMR gene

Chromosomal DNA isolated from dye degrading organism was used as the template for PCR.

Polymerase Chain Reaction

In molecular biology, the polymerase chain reaction (PCR) is a technique to amplify a single or few copies of a piece of DNA across several orders of magnitude, generating thousands to millions of copies of a particular DNA sequence. The method relies on thermal cycling, consisting of cycles of repeated heating and cooling of the reaction for DNA melting and enzymatic replication of the DNA. Complementary to the target region along with a DNA polymerase (after which the method is named) are key components to enable selective and repeated amplification.. PCR can be extensively modified to perform a wide array of genetic manipulations.

Results and Discussion

Five grams of soil sample was inoculated in 100ml of sterilized Nutrient Broth and were incubated at 28^0C on incubation shaker for 72h and 160rpm until growth appeared. 100µl of suspension was spread over Nutrient Agar plates along with a dye (bromophenol blue, crystal violet or malachite green) with concentrations ranging from 20-80µM. The petri plates are incubated for 48 hrs at 37^0c. Bacterial isolates from the dye contaminated soil were found to be capable of growing in media containing the triphenyl methane dye. The colony which has shown maximum clear zones (i.e. maximum decolorization) in that particular dye was observed.

Decolorization assay

The decolorization assay was done for the three dyes and the OD values of the incubated nutrient broth samples were recorded show as Figure 1. Pure culture of isolated strains tested individually can decolorize crystal violet (CV), bromophenol blue (BB) and malachite green (MG) with an efficient rate. Greater decolorization was reached at 24 h as indicated in Table 1. Interestingly, decolorization was more rapid (85%) for BB compared to MG (77%) and CV (75%) at 24 h.

Decolourization assay indicated that maximum decolourisation occurred at lower concentrations (20µM). The highest decoloration was seen in the case of bromophenol blue i.e. 85%. In case of crystal violet maximum decolorization (96.3%) has been achieved at 29.38 ppm of crystal violet by fungal system *Cyathusbulleri* where as with bacterial system *Pseudomonas pseudumallei*13 NA and *Rhodotorulaerubra*, maximum of 20.4 and 10 ppm of crystal violet were decolorized effectively to the extent of 96 and 99 % respectively. As compared to crystal

8

violet higher concentration of malachite green was decolorized by the same amount of cell mass, which may be due to the difference in the structure of both the dyes. The rate of decolorization of malachite green by *Phanerochaetechrysosporium*was found to be higher than crystal violet. Since malachite green has two dimethyl groups in two side chains where as crystal violet is having three dimethyl groups in three side chains which may be the reason that decolorization of more substituted triphenyl methane dyes took longer time.

Enzyme activity assay

Since the dyes are triphenyl methane dyes and the organism is able to reduce the dye to a maximum extent at 20 μmol, therefore triphenyl methane reductase enzyme assay was done at 20 μmol concentration. The enzyme activity was calculated based on the OD values obtained at 600 nm. The results of the enzyme assay for the three dyes at 20 μmol concentration is given in Table-2.

TMR enzyme activity was found to be the highest in bromophenol blue (0.0046 μ/min/ml), followed by malachite green (0.0029 μ/min/ml) and the least was seen with crystal violet (0.0014 μ/min/ml).

Characterization of the bacterial isolates

The three isolates D1, D2 and D3 were characterized based on microscopic and biochemical tests. The results are given in Table-3.

Colony Morphology

The colonies that decolorized the dye were large, low convex, rough and oval.

Gram staining:

The organisms which have shown clear zones (i.e. maximum decolourisation) were picked and maintained pure cultures. These isolates were analyzed for their gram nature. Gram staining revealed the isolated organism to be gram negative bacilli

Biochemical Tests

The culture was positive for gelatin utilization, oxidase positive and negative for catalase test, urease test.

From the above IMViC and biochemical tests, the three isolates isolated organism was identified as belonging to *Pseudomonas species*.

DNA isolation:

From the pure culture of Pseudomonas sp, DNA was isolated by non enzymatic method. Approximately 0.07µg of DNA was obtained. The DNA was used for amplification of TMR gene in PCR machine.

Conclusion

The present study was carried out to examine the microbial degradation of textile dyes like bromophenol blue, crystal violet, and malachite green by the organism isolated from soil. The isolates have shown positive results for dye degradation as was indicated by the change and disappearance of colour of the dye from the dye-containing media of the petri plates. The degradation/decolorization of the tested dye might be due to the production of extracellular enzymes which was quantified and was found to be in the range of 0.0012- 0.0046 (u/min/ml).TMR gene was isolated and amplified by PCR.

References

1. Alinsafi A., Motta M. and Benhammou A.; 2005. Effect of variability on the treatment of textile dyeing wastewater by activated sludge. Dyes and Pigments. : 69(1-2):31–39.

2. Bell J., Buckley C.A. and Stuckey D.C.; 2000. Treatment and decolorization of dyes in an anaerobic baffled reactor. J Environ Eng Div ASCE 126:1026 –32.

3. Cariell C.M., Barclay S.J. and Bukley C.A.; 1996. Treatment of exhausted reactive dye bath effluent using anaerobic digestion: laboratory and full scale trials. Water S.A. 21, 225-233.

4. Daneshvar N., Ayazloo M. and Khataee A.R.; 2007. Biological decolorization of dye solution containing Malachite Green by microalgae *Cosmarium* sp. Bioresource Technology.: 98(6):1176–1182.

5. Lacalle M.l., Villaverde S., Fdz-Polanco F. and Garcia-Encina P.A. ; 2001. Combined anaerobic/aerobic (UASB+UBAF) system for organic matter and nitrogen removal from a high strenth industrial wastewater. Water Sci. Technol. 44: 255-62.

6. Oxspring D.A., McMullan G., Smyth W F. and Marchant R.; 1996. Decolourisation and metabolism of the reactive textile dye, remazol black B, by an immobilized microbial consortium. Biotech Lett 18:527 –30.

7. Robinson T., McMullan G., Marchant R. and Nigam P.; 2001. Remediation of dyes in textile effluent: a critical review on current treatment technologies with a proposed alternative. Bioresource Technology. 77(3):247–255.

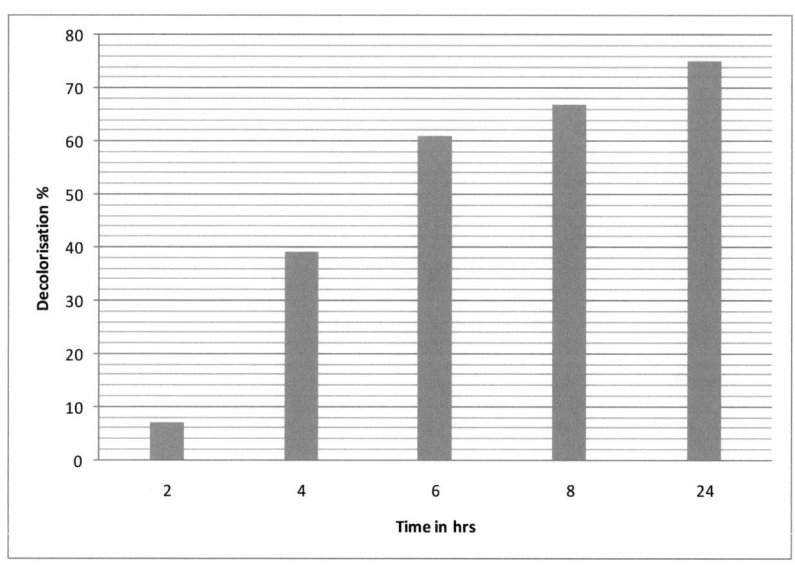

Figure: 1 Results of decolorization assay of malachite green, crystal violet and bromophenol blue

Table-1 Results of decolorization assay of malachite green, crystal violet and bromophenol blue

TYPE OF DYE	2hrs	4hrs	6hrs	8hrs	24hrs
CV	7	39	61	67	75
MG	0.6	20	53	69	77
BB	6	25	58	68	85

Table-2   TMR   Enzyme   activities   at   different   concentrations   of   dye

Dye at 20 μmol concentration	Optical Density at 600nm	Enzyme Activity(μ/min/ml)
Bromophenol blue	1.44	0.0046
Crystal violet	0.28	0.0012
Malachite green	0.44	0.0029

Table-3 Results of the IMVIc Test Isolates

S. No	Gram Nature	Indole Production	Methyl Red Reduction	Voges Proskauer	Citrate Utilization
D 1	Gram negative	-ve	-ve	-ve	+ve
D2	Gram negative	-ve	-ve	-ve	+ve
D 3	Gram negative	-ve	-ve	-ve	+ve